See It Grow

STRAWBERRY

by E. Merwin

Consultant: Karen C. Hall, PhD
Botanist
Eugene, Oregon

BEARPORT PUBLISHING

New York, New York

Credits

Cover, © Igor Marx/Shutterstock, © Igor Borodin/Shutterstock, and © Anton-Burakov/Shutterstock; Title Page, © Tim UR/Shutterstock; TOC, © anna1311/iStock and © 5 second Studio/Shutterstock; 4, © Yakov Oskanov/Shutterstock; 5, © Goce Risteski/Alamy; 6, © Nicole_P_Pix2017/Shutterstock; 7, © elf911/iStock, © Nicole_P_Pix2017/Shutterstock, © Elena Kitch/Shutterstock, and © ifong/Shutterstock; 8, © Elena Kitch/Shutterstock, © xtrekx/iStock, © Eric Isselee/Shutterstock, © Sony Ho/Shutterstock, © ifong/Shutterstock, and © irin-k/Shutterstock; 9, © Anna Mikheeva/Shutterstock; 10T, © Veronkia Polbina/Shutterstock; 10B, © Alena Brozova/Shutterstock; 11T, © Alena Brozova/Shutterstock; 11B, © Artem Kutsenko/Shutterstock; 12, © Wicki58/iStock; 13, © Tiplyashina Evgeniya/Shutterstock; 14–15, © Volodymyr Nikitenko/Shutterstock; 16, © Anest/Shutterstock; 17, © Elena Masiutkina/Shutterstock; 18, © Elena Kitch/Shutterstock, © xtrekx/iStock, © Eric Isselee/Shutterstock, © Sony Ho/Shutterstock, © Anest/Shutterstock, © ifong/Shutterstock; 19T, © EduardSV/Shutterstock; 19B, © ER_09/Shutterstock; 20, © Dennis Kuvaev/Shutterstock; 21L, © chuckcollier/iStock; 21RT, © Alter-ego/Shutterstock; 21RB, © Oliver Hoffmann/Shutterstock; 22, © Teshima Yasuhito/Cover Images/Newscom; 23 (T to B), © Anest/Shutterstock, © Zoom Team/Shutterstock, © manfredxy/Shutterstock, and © Madlen/Shutterstock; 24, © Maks Narodenko/Shutterstock; Back Cover, © Natalia K/Shutterstock.

Publisher: Kenn Goin
Senior Editor: Joyce Tavolacci
Creative Director: Spencer Brinker
Design: Debrah Kaiser
Photo Researcher: Thomas Persano

Library of Congress Cataloging-in-Publication Data

Names: Merwin, E., author.
Title: Strawberry / by E. Merwin.
Description: New York, New York : Bearport Publishing Company, 2019. |
Series: See it grow | Includes bibliographical references and index.
Identifiers: LCCN 2018009261 (print) | LCCN 2018011408 (ebook) |
ISBN 9781684027224 (ebook) | ISBN 9781684026760 (library)
Subjects: LCSH: Strawberries—Juvenile literature.
Classification: LCC SB385 (ebook) | LCC SB385 .M58 2018 (print) |
DDC 634/.75—dc23
LC record available at https://lccn.loc.gov/2018009261

For more information, write to Bearport Publishing Company, Inc., 45 West 21st Street, Suite 3B, New York, New York 10010. Printed in the United States of America.

10 9 8 7 6 5 4 3 2 1

Contents

Strawberry

Strawberries are yummy!

They are red, sweet, and juicy!

How did they get that way?

Strawberries are one of the most popular fruits!

Strawberry plants begin as tiny seeds.

A strawberry seed is as small as a flea!

After a seed is planted, it begins to grow.

First, it sends roots down into the **soil**.

Then, green leaves start to appear.

Over time, the strawberry plant grows more leaves.

The plant's base is called a crown.

From the crown, small white flowers grow.

Bees land on the flowers.

Insects, such as bees, help **pollinate** the flowers.

The flowers turn into pea-sized green berries.

As they get bigger, the berries change from green to white.

Soon, the strawberries turn pinkish in color.

Then, they become bright red!

It takes about one month for a berry to grow and ripen.

Now, the strawberries are juicy and ripe.

Farmers pick them by hand.

The berries are soft and easy to squish.

Most of the world's strawberries are grown in California.

Strawberry plants can grow another way, too!

Fully grown plants shoot out long stems, or runners.

These runners can make new plants!

A single strawberry plant can send out 30 to 50 runners.

Little green **nodes** grow on each runner.

When a node touches the earth, new roots grow.

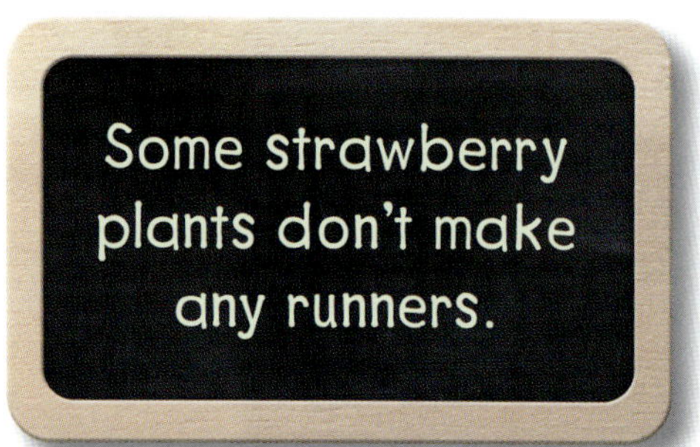

This new plant is called a daughter plant.

The mother plant passes **nutrients** to the daughter plant.

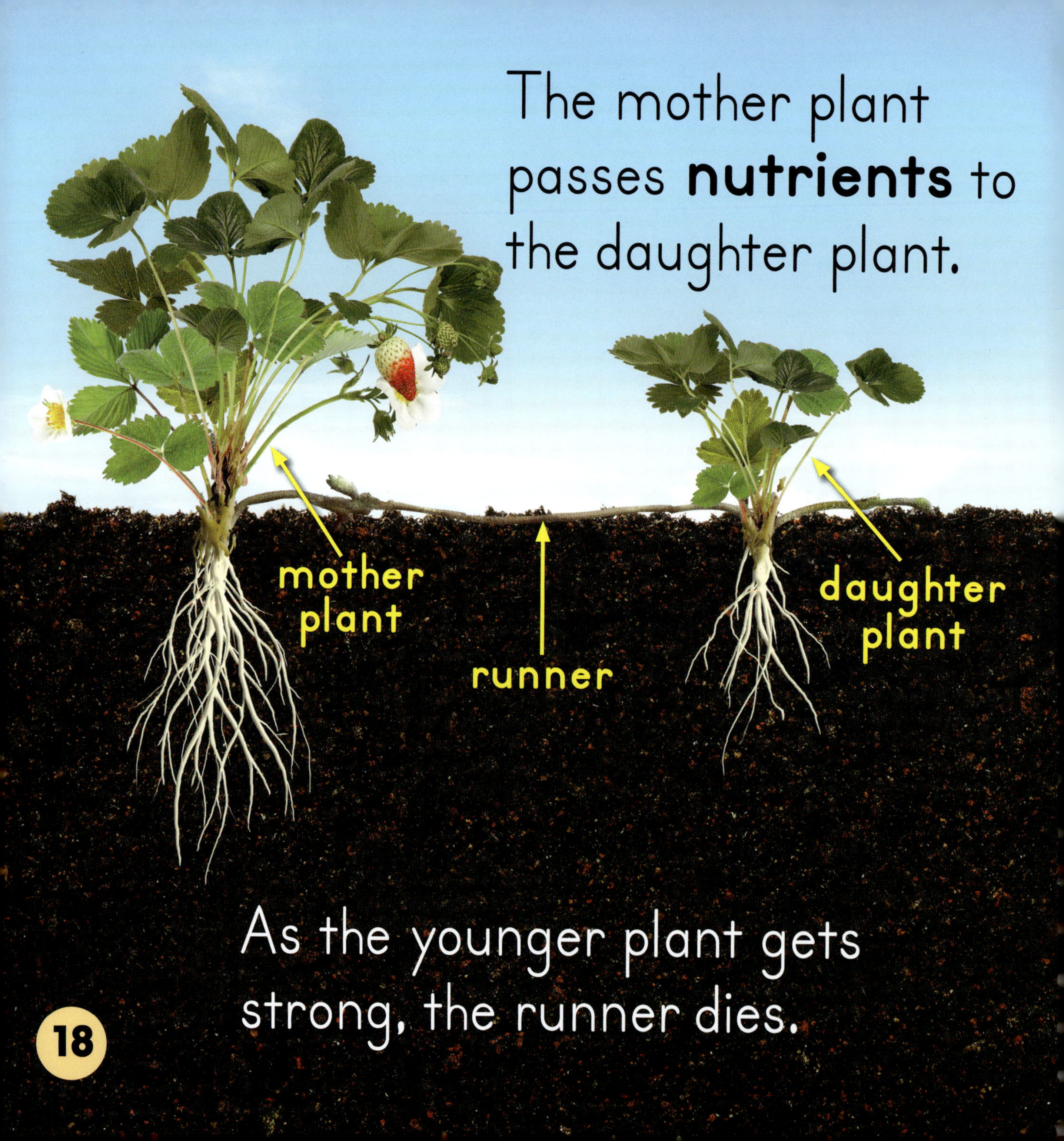

As the younger plant gets strong, the runner dies.

Now, the daughter plant grows on its own!

The new plant will soon grow its own berries.

Most of the strawberries we eat come from daughter plants.

What are those dark specks on the strawberry's skin?

Each one is a tiny seed!

People love gobbling up fresh strawberries. The berries can also be made into jam, ice cream, and cakes.

Strawberry Facts

- The strawberry is a symbol of love. Why? It's red and shaped like a heart!
- It takes about 65 strawberries to make a strawberry pie.
- The White Jewel is a special white strawberry grown in Japan. It costs about $10 for each berry!

- Strawberries are good for your body. They're packed with vitamin C.

Glossary

nodes (NOHDZ) the parts of a plant stem from which leaves and roots can grow

nutrients (NOO-tree-uhnts) substances needed by plants to grow and stay healthy

pollinate (POL-uh-nayt) to carry pollen from one flower to another, which allows the second flower to make fruit and seeds

soil (SOYL) dirt in which plants grow

Index

Read More

Christensen, Bonnie. *Plant a Little Seed.* New York: Roaring Brook (2012).

Schuh, Mari. *From Seed to Strawberry.* New York: Lerner (2016).

Learn More Online

To learn more about strawberries, visit
www.bearportpublishing.com/SeeItGrow

About the Author

E. Merwin is a writer who much admires nature—and the way all things grow!